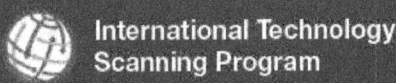

International Technology Scanning Program

August 2012

Developing Multilevel Memorandums of Understanding With Utility Companies

SPONSORED BY:

U.S. Department of Transportation
Federal Highway Administration

IN COOPERATION WITH:

American Association of State Highway and Transportation Officials

National Cooperative Highway Research Program

NOTICE

The Federal Highway Administration provides high-quality information to serve Government, industry, and the public in a manner that promotes public understanding. Standards and policies are used to ensure and maximize the quality, objectivity, utility, and integrity of its information. FHWA periodically reviews quality issues and adjusts its programs and processes to ensure continuous quality improvement.

Technical Report Documentation Page

1. Report No. FHWA-PL-12-025	2. Government Accession No.	3. Recipient's Catalog No.	
4. Title and Subtitle Developing Multilevel Memorandums of Understanding With Utility Companies		5. Report Date August 2012	
		6. Performing Organization Code	
7. Author(s) C. Paul Scott, P.E., Cardno TBE		8. Performing Organization Report No.	
9. Performing Organization Name and Address American Trade Initiatives 3 Fairfield Court Stafford, VA 22554-1716		10. Work Unit No. (TRAIS)	
		11. Contract or Grant No. DTFH61-10-C-00027	
12. Sponsoring Agency Name and Address Office of International Programs Federal Highway Administration U.S. Department of Transportation American Association of State Highway and Transportation Officials		13. Type of Report and Period Covered	
		14. Sponsoring Agency Code	
15. Supplementary Notes FHWA COTR: Hana Maier, Office of International Programs			
16. Abstract The Federal Highway Administration, American Association of State Highway and Transportation Officials, and National Cooperative Highway Research Program sponsored a scanning study of innovative right-of-way and utility practices that might be applicable in the United States. The scan team found that two Australian states, Queensland and New South Wales, have developed memorandums of understanding (MOUs) and similar agreements with utility companies to facilitate cooperation and optimize relationships between transportation agencies and utilities. This report provides the results of a survey of State highway agencies in the United States on partnering agreements with utility companies and whether any have agreements similar to the Australian MOUs. It also provides a step-by-step approach for developing an Australian-type MOU in the United States, a sample MOU template, and a sample conflict resolution matrix for handling differences that may arise.			
17. Key Words Master agreement, memorandum of understanding, partnering agreement, right-of-way, utility relocation		18. Distribution Statement No restrictions. This document is available to the public from the: Office of International Programs, FHWA-HPIP, Room 3325, U.S. Department of Transportation, Washington, DC 20590 international@fhwa.dot.gov www.international.fhwa.dot.gov	
19. Security Classify. (of this report) Unclassified	20. Security Classify. (of this page) Unclassified	21. No. of Pages 47	22. Price Free

Form DOT F 1700.7 (8-72)   Reproduction of completed page authorized

Contents

Executive Summary 1

Chapter 1:
Introduction and Study Scope 3

Chapter 2:
Australian Memorandums of Understanding 5

Chapter 3:
U.S. Partnering Agreements 10

Chapter 4:
Comparison of Australian MOUs and U.S. Partnering Agreements 17

Chapter 5:
Findings and Recommendations 19

Chapter 6:
Step-by-Step Suggestions for Developing an MOU in the United States 22

References and Resources 28

Appendix A:
Summary of Australian Memorandums of Understanding 29

Appendix B:
Summary of Responses to State Department of Transportation Survey 30

Appendix C:
Summary of U.S. Partnering Agreements 40

Appendix D:
Sample Conflict Resolution Matrix 45

Executive Summary

In September 2008, an international scan team visited Australia and Canada to learn about innovative right-of-way and utility practices that might be applicable in the United States. The scan team found that two Australian states, Queensland and New South Wales, have developed memorandums of understanding (MOUs) and similar agreements with major utility companies to facilitate the cooperation and coordination process and to optimize the relationship between transportation agency and utility interests.

The basic Australian MOU structure includes a high-level MOU developed by upper managers that sets general principles and the intent of both parties to cooperate; several midlevel MOUs developed by middle managers that define roles and responsibilities, specifications, and general procedures to resolve high-priority conflict situations; and project-specific contract-level agreements that detail specific provisions not addressed in the higher level MOUs.

Many States in the United States have partnering agreements with utility companies. To determine if any State partnering agreements were similar to the Australian MOUs, a survey was sent to statewide utilities managers in 50 States, Puerto Rico, and the District of Columbia. The survey asked if they had any such partnering agreements. Forty-six States, Puerto Rico, and the District of Columbia responded to the survey. Of these 48 respondents, only the Maine Department of Transportation (DOT) responded that it had partnering agreements similar to the Australians. Twenty State DOTs, including Maine, provided copies of their partnering agreements. These partnering agreements are summarized in this report. Ten State DOTs expressed interest in participating in a pilot project to develop and evaluate MOUs similar to the Australian MOUs.

A comparison of Australian and American partnering agreements revealed the following:

- Partnering agreements are similar in their primary purpose, which is to improve working conditions (coordination, cooperation, communication, and commitment) with utility companies.

- Partnering agreements are different in levels of development. The Australian MOUs were developed at a high organizational level, while partnering agreements in the United States were generally developed at lower organizational levels.

- Partnering agreements are different in reimbursement and access requirements. Utility companies are reimbursed 100 percent for relocations in Australia and some have unlimited access to the right-of way. DOTs in the United States have total control of the highway right-of-way, and the majority reimburse utility companies for required relocations only if the utility companies have prior property rights.

- Partnering agreements are different in stated conditions for noncompliance. The Australians have generally included a shared-risk process in which participants are responsible for a shared portion of extra costs if they do not do what they have agreed to do. In the United States, partnering agreements are often considered nonbinding and provisions, if any, for noncompliance only involve termination clauses.

Other similarities and differences exist, but these are the most relevant to any efforts in the United States to promote the Australian MOU concept.

MOUs similar to those used in Australia might be a considerable help in advancing highway projects to an earlier and less contentious completion. However, implementation in the United States will most likely occur only if statewide utility managers champion the effort and high-level officials of State DOTs and utility companies commit to cooperate, coordinate, and communicate. This report contains a step-by-step approach for developing an Australian-type MOU in the United States, a sample MOU template, and a sample conflict resolution matrix for handling differences that may arise.

CHAPTER 1

Introduction and Study Scope

Introduction

In September 2008, an international scan team visited Australia and Canada to learn about innovative right-of-way and utility practices that might be applicable in the United States. Findings from the scanning study are documented in the team's report, *Streamlining and Integrating Right-of-Way and Utility Processes With Planning, Environmental, and Design Processes in Australia and Canada* (FHWA-PL-09-011). Some of the report findings are included in this report in discussions of Australian MOUs.

The scan team found that two Australian states, Queensland and New South Wales, have developed memorandums of understanding (MOUs) and similar agreements with major utility companies to facilitate the cooperation and coordination process and to optimize the relationship between transportation agency and utility interests.

The basic Australian MOU structure includes a high-level MOU developed by upper managers that sets general principles and the intent of both parties to cooperate; several midlevel MOUs developed by middle managers that define roles and responsibilities, specifications, and general procedures to resolve high-priority conflict situations; and project-specific contract-level agreements that detail provisions not addressed in the higher level MOUs.

The Department of Main Roads (DMR) in Queensland has two MOUs in place—one with Energex, a major electric utility, that focuses on specifications, utility relocation estimates, and prenotices and one with Telstra, a major telecommunications utility, that focuses on asset protection, access, and contracts. Two more MOUs are in the development phase.

The Roads and Traffic Authority (RTA) in New South Wales has an MOU in place with the Sydney Water Corporation that establishes a framework covering issues such as cost distribution, information sharing, strategic planning, project management, and dispute resolution. RTA is working on a similar MOU with Energy Australia.

The multilevel MOU concept has not been used in the United States, but many States do have partnering agreements with utility companies. However, Australian MOUs are generally more elaborate and stringent than U.S. partnering agreements. This may be because Federal, State, and local laws, regulations, and policies in the United States govern the relocation and accommodation of utilities on highway right-of-way, but a similar concept does not appear to exist in Australia. The probable reason is that in Australia states normally reimburse utility interests in full for relocating utility facilities because historically most were government entities. In recent years, the Australian utility industry has undergone deregulation and a large percentage of utility interests are now in private hands, but the policy for reimbursing utility relocations has continued.

Study Scope

The objective of this study was to establish a recommended step-by-step process and a template for roles and responsibilities of multiple parties that might want to develop and use multilevel MOUs. This was accomplished through the following:

- Conducting a literature review of information obtained from the Australian states (Queensland and New South Wales) that use multilevel MOUs. The literature review included a determination of how the MOUs were drafted, the language used, the levels of organizational leadership that developed and administer the MOUs, a description of the Australians' experiences with barriers and successes in implementing the MOUs, and specific details about the use of the MOUs.

- Conducting an e-mail and telephone survey of State departments of transportation (DOTs) in the United States to determine which have MOUs, master agreements, or other partnering agreements similar to the Australian model and to determine the utilities with which these agreements exist. Statewide utility managers were contacted in 50 States, the District of Columbia, and Puerto Rico. Forty-six States, the District of Columbia, and Puerto Rico responded to the survey.

- Preparing a final report that includes a comparison of the Australian and American models, step-by-step recommendations for developing an MOU in the United States, presentation of a template for an Americanized version of the multiple-level MOU for appropriate use by U.S. utility companies and transportation agencies, and identification of a minimum of three States willing to experiment with using the multiple-level MOU approach and record the lessons learned after a trial period.

CHAPTER 2

Australian Memorandums of Understanding

Two states in Australia, Queensland and New South Wales, are exploring a variety of MOUs and similar agreements with utility agencies to facilitate the cooperation and coordination process and to optimize the relationship between transportation agency and utility interests. In a typical situation, a high-level MOU sets general principles and the intent of both parties to cooperate. To ensure that the MOU is a living document, it may include attachments and other agreements that discuss specific issues, such as standards, specifications, and general procedures for resolving conflict situations. Typically, technical personnel from both organizations prepare these midlevel documents. There might also be contract-level agreement details and specific provisions not addressed in the higher level MOU.

DMR in Queensland has two MOUs in place—one with Energex, a major electric utility, that focuses on specifications, relocation estimates, and prenotices and one with Telstra, a major telecommunications utility, that focuses on asset protection, access, and contracts. Two more MOUs are in the development phase.

RTA in New South Wales has an MOU in place with the Sydney Water Corporation that establishes a framework covering issues such as cost distribution, information sharing, strategic planning, project management, and dispute resolution. RTA is working on a similar MOU with Energy Australia.

The Australian MOUs are more elaborate and stringent than the MOUs, master agreements, and other partnering agreements used in the United States. As opposed to the United States where Federal and State laws, regulations, policies, and rules govern the relocation, reimbursement, and accommodation of utilities on highway rights-of-way, a similar concept does not exist in Australia, which could explain in part the need for more comprehensive MOUs. Also, telecommunications companies in Australia have unlimited access to highway rights-of-way, while State and local highway authorities in the United States have complete control of utilities on right-of-way that they manage.

A summary of the Queensland and New South Wales memorandums of understanding are in Appendix A. More details on these MOUs are in the following sections.

Queensland

The multilevel MOUs in Queensland include a high-level MOU setting general principles and the intent of both parties to cooperate; midlevel MOUs defining roles and responsibilities, standards, specifications, and general procedures for the resolution of conflict situations; and contract-level agreements detailing specific provisions that the higher level MOUs do not address. To date, only the high-level MOUs have been put in place and the midlevel MOUs are being developed.

Main Roads–Energex MOU

According to its Web site, Energex is one of the Australia's largest and fastest-growing organizations with more than 3,500 staff working in a range of roles to supply electricity to about 2.8 million people in South East Queensland. In 2002, Energex officials helped develop and entered into an MOU with the Queensland DMR.

The first steps in the process of putting together this MOU involved bringing together high-level officials to identify and prioritize the issues and concerns that impacted working operations of both organizations. Although high-level officials from both Energex and DMR were involved in developing the MOU, they appointed a steering committee to monitor the work and joint working groups to do the day-to-day work.

Energex and DMR officials deemed development of an MOU necessary because poor working relationships were hampering the operations of both organizations. Several Energex activities significantly impacted DMR's operations, including the following:

- DMR required Energex to install cables at least 1.2 meters (m) deep under road pavements to eliminate the possible need for expensive reimbursable utility relocations associated with future road work. Energex, however, did not want to install cables more than 0.9 m deep because any deeper would not allow for dissipation of heat from high-energy cables (the deeper they are laid, the hotter they get, which may contribute to cable failure).

- Energex's contractors had done work in DMR's corridors with little apparent regard for environmental damage. There had been cases in which protected species were identified and Energex was notified of their existence, but the contractor nonetheless cleared the corridor without regard for the protected species.

- On many occasions, utility poles were erected well inside the defined clear zones, creating safety and liability concerns.

- Delays on projects were a common occurrence. At times DMR's notifications and responses to Energex were unreasonable and Energex gave higher priority to its own program or otherwise delayed projects because of lack of available resources or personnel.

- Energex's costs to relocate its assets on DMR's projects were at times based on rough estimates because the utility knew it would be reimbursed for the actual costs at the time of construction, regardless of what they turned out to be. In some cases, actual costs were double the original estimates, impacting DMR's total project cost.

- Energex frequently installed new equipment in the road corridor regardless of plans for future road widening works and expected DMR to bear relocation costs at the time of road works.

Energex also had issues with DMR. For example, there was a significant degree of inconsistency among the DMR's regional offices that dealt with utilities. Approaches, processes, systems, priorities, and decision criteria differed to an extent that both organizations were frustrated. As a result of all the issues, relationships between DMR and Energex had not been good.

When work began on the MOU, two principal areas were selected for addressing issues between DMR and Energex:

- Construction and maintenance operations

- Corridor planning, access, and infrastructure design

The steering committee identified primary issues for each of the principal areas. Construction and maintenance operations issues included the following:

- Rationalizing and sequencing to better coordinate Energex and DMR activities and establish commitments to set timeframes for project delivery

- Timing to ensure that adequate notice would be given to Energex to enable it to relocate its facilities in a timely manner

- Future costs to address the longstanding issue of DMR being expected to pay to relocate conflicting Energex plant in all cases

- Details addressing road safety considerations for locating Energex's power poles in the highway right-of-way

Corridor planning, access, and infrastructure design issues included the following:

- Strategic planning to determine benefits likely to be gained from sharing information and working together

- Development of planning guidelines for installing new Energex plant in the road corridor to avoid future conflicts and minimize future relocation costs

- Incorporation of future cost considerations into all policies, agreements, processes, and decisions on corridor planning access and infrastructure design

- Development of a process to consider overhead and underground installations in future planning and design

These and other primary issues were resolved through agreed-on outcomes and documented in signed agreements. All outcomes were signed by the respective organizations' chief executive officers or the steering committee and implemented in DMR's and Energex's working operations.

Energex and DMR are now working on second-level priorities. These include the following:

- Development of a new system to better manage utility assets on DMR projects

- Development of a system to identify power pole locations that provide the greatest risk potential to motorists and help reduce road fatalities and serious injury crashes

- Land access notifications made up of an agreement document, general technical conditions, and an accommodation process

- Development of a system involving proposed Energex trenching works across road pavement for new underground installations, particularly for the use of flowable fill and lean-mix backfill

- Estimating with pictures to help DMR quickly identify Energex's assets and develop a reasonable cost estimate for relocations necessitated by highway construction

Main Roads–Telstra MOU

According to its Web page, Telstra is Australia's leading telecommunications and information services company, with one of the best known brands in the country. It offers a full range of products and services and competes in all telecommunications markets throughout Australia, providing 9.2 million fixed-line services and 9.7 million mobile services. In 2005, Telstra officials helped develop and entered into an MOU with DMR.

The first steps in the process of putting together this MOU involved bringing together high-level officials to identify and prioritize the issues and concerns that impacted working operations of both organizations. Although high-level officials from Telstra and DMR were involved in developing the MOU, they appointed a steering committee to monitor the work and working groups to do the day-to-day work.

When work to develop an MOU began, three main areas causing difficulties for both Telstra and DMR were identified:

- Asset protection

- Access to roads
- Contracts

Primary issues were identified for each of the principal areas. Asset protection issues included the following:

- Development of new asset protection procedures for telecommunications assets, including roles, responsibilities, cost minimization, variations at the construction stage, and a consistent statewide process among all DMR regional offices

- Development of guidelines for protecting and installing new telecommunications plant in the road corridor to avoid conflicts with future road works, minimize and avoid future costs for relocations, and minimize disruptions to Telstra's service. These guidelines will provide support during project development, in determining roadway alignment and road widening options, and for projects in the development and implementation stages.

- Development of guidelines to address future cost-sharing responsibilities between DMR and Telstra for installing Telstra's assets identified as being in conflict with future upgrading works on state-controlled roads

Access to roads issues included the following:

- Development of a land access agreement with procedures, decision criteria, and standards on notifying and conducting maintenance work and installations by Telstra on DMR-controlled roads, as required by the Telecommunications Act of 1997

Contract issues included the following:

- Development of an asset protection contract formalizing the relationship between DMR and Telstra for delivering the asset protection works

- Development of a procedure to minimize damage to telecommunications plant and other services by the contractor during roadway construction

A memorandum of understanding was developed identifying the primary priorities group, with targeted agreed-on outcomes, the benefits to be gained, and the degree of complexity to achieve these goals. All outcomes were signed by either the organizations' chief executive officers or the steering committee and implemented in DMR's and Telstra's working operations.

Benefits

DMR, Energex, and Telstra believe that their high-level partnering agreements have improved working relationships and trust. Roles and responsibilities between the agencies are more clearly defined. Agreed-on outcomes and new processes have provided win-win solutions to problems. Other recognized benefits include enhanced sharing of confidential information, better cost estimating for projects, and major time and cost savings for both organizations.

Compliance

Parties to the MOUs are responsible for meeting the terms of the agreement. If they do not, it is anticipated that the situation will be handled in a shared-risk process in which the participants pay portions of the costs of not doing what they were required to do.

New South Wales

RTA and the Sydney Water Corporation entered into an MOU to promote cooperation and good practices in the interaction between both parties on adjustments and relocations necessitated by highway construction. RTA is working on a similar MOU with Energy Australia.

RTA—Sydney Water MOU

According to its Web site, Sydney Water is Australia's largest water utility with more than 3,200 staff and an operations area covering 12,700 square kilometers. A corporation wholly owned by the New South Wales government, it has three equal objectives: protect public health, protect the environment, and be a successful business.

The MOU is based on a model agreement developed by the New South Wales' Streets Opening Conference (NSW SOC) in 1999. The NSW SOC started in Sydney in 1909 as a focal point for discussing common transportation and utility issues. The conference is not an event, but rather is a voluntary association of member organizations. Its objectives include establishing roadside allocations and recommended practices for providing utility services; fostering coordination; encouraging the use of codes and practices for excavation, backfilling, and roadway reconstruction; and minimizing the impact of excavations. Membership includes utility owners, local government and road authorities, light rail operators, other government agencies, consultants, and other groups with an interest in utility issues.

Over the years, the NSW SOC has undertaken major initiatives, such as the model agreement for strategic alliances between utilities and road authorities; guides to codes and industry practices; a specification on clearing, excavating, backfilling, and other activities associated with utility installations; a pilot training course to improve the understanding of plans and identification of facility components by technicians and contractors; and a dial-before-you-dig referral system for information on underground utility installations.

The RTA–Sydney Water MOU was founded on the premise that coming to an understanding on cooperative processes provides the key to better meeting the objectives of both agencies. As with the Queensland MOUs, officials from RTA and the water company identified primary issues and agreed on outcomes.

Issues covered in the RTA–Sydney Water MOU are as follows:

- Cost distribution
- Information sharing
- Strategic planning
- Project management
- Dispute resolution

The MOU includes case study scenarios that describe typical situations and provide additional information, such as agreement and cost distribution.

The MOU was signed by chief executive officers from both organizations and incorporated in working operations. Both RTA and Sydney Water hope their partnering agreement will improve working relationships. No process is in place to measure the results, and the MOU has been in place only for a short time. Early signs indicate, however, that interactions between the two entities are much improved.

CHAPTER 3

U.S. Partnering Agreements

Historically in the United States, it has been in the public interest for utility facilities to use and occupy the rights-of-way of public roads and streets. Highway and utility officials have two broad areas of concern about the practice of jointly using a common right-of-way. First is the cost of relocating, replacing, or adjusting utility facilities that fall in the path of proposed highway improvement projects, and second is the installation of utility facilities along or across highway rights-of-way and the way they occupy and jointly use those rights-of-way.

The Federal Highway Administration's (FHWA) *Utility Relocation and Accommodation: A History of Federal Policy Under the Federal-Aid Highway Program* provides an early history of utility accommodation on highway right-of-way. In 1946, the Public Roads Administration commissioner issued the first all-inclusive detailed instructions on utility adjustments in a single document to Public Roads Administration field offices and State highway departments. These instructions and subsequent amendments are now contained in Title 23, Code of Federal Regulations, Part 645 (23 CFR 645). In retrospect, the 1946 policy statement can be viewed as a remarkable document. Many of its basic provisions have withstood the test of time and operations for almost 66 years under the largest public works program ever undertaken, and most are valid today except for minor updates and additions stemming from new Federal laws. Even though these requirements apply to federally funded projects, all State departments of transportation (DOTs) have adopted them for State and local work.

Section 113 of 23 CFR 645 requires utility companies and State DOTs to agree in writing on their separate responsibilities for financing and accomplishing relocation work necessitated by highway construction. To comply with this Federal requirement and State requirements, which may be more stringent than Federal requirements, some State DOTs have developed various forms of partnering agreements. While FHWA has had requirements in place since 1946 for adjusting and relocating utilities on highway projects and only requires a separate agreement when reimbursement is involved, the Australian states have had a greater need for partnering agreements because they have had to rely on individual agreements containing all needed requirements for every project.

FHWA's *Program Guide: Utility Relocation and Accommodation on Federal-Aid Highway Projects* (FHWA-IF-03-014) further specifies that the agreement between a utility and a State DOT that describes separate responsibilities for financing and accomplishing relocation and adjustment work may be either a master agreement on an areawide or statewide basis or individual agreements for utility work on a case-by-case or project basis. To comply with Federal and State requirements, some States have developed master agreements, master contracts, MOUs, standard reimbursement agreements, and other such partnering agreements.

To determine if any State DOTs had partnering agreements with utility companies similar to the Australian MOUs, a survey was sent to State DOT statewide utilities managers. Forty-six States, Puerto Rico, and the District of Columbia responded to the survey. Of these 48 respondents, only the Maine DOT had partnering agreements similar to the Australians. More details provided by State DOTs are in Appendix B (Summary of Responses to State DOT Survey) and Appendix C (Summary of U.S. Partnering Agreements).

Twenty respondents sent copies of their partnering agreements, summarized in this chapter. Other State

DOTs probably have partnering agreements similar to those provided, but the following agreements provide a general overview of partnering agreements nationwide.

Alaska

The Alaska DOT has an MOU that it uses in conjunction with individual construction projects. A copy of a project-specific MOU with Enstar Natural Gas Company was provided in response to the survey. The MOU establishes the basis for reimbursement to utility companies for preliminary engineering and/or adjustment and relocation of utility facilities when necessitated by highway construction. These project-specific MOUs are signed by the agency's utility section chief and designated utility representatives.

Arizona

The Arizona DOT has a standard utility master agreement that it uses in conjunction with individual construction projects. A copy of a utility master agreement with the Salt River Project (SRP) Agricultural Improvement and Power District, which provides electricity to nearly 930,000 retail customers in the Phoenix area, was provided in response to the survey. The purpose of the agreement is to establish requirements for relocating, adjusting, or rearranging certain existing facilities belonging to utility companies to facilitate construction of highway projects by the Arizona DOT. Provisions in the agreement cover (1) construction requirements (staking, removal of abandoned facilities, disposal sites, damage to facilities, ownership of facilities, transfer of land interests, land transfer, prior rights, right-of-way delays, construction right-of-way, permits, traffic, start and completion date, subcontracting, Blue Stake one-call notification system, acceptance date), (2) construction schedules (schedule changes, failure to meet schedule, payments for interference, changes in scope of work, cost increases, submission of invoices, payment of invoices, reimbursements, expeditious payments and remedy for late payment, nonwaiver), (3) liability, (4) statutorily mandated items, and (5) miscellaneous conditions. The agreement was signed by the Arizona DOT engineer and manager of the utility section and the SRP department manager.

California

The California DOT (Caltrans) has a freeway master contract in place with eight major utility companies and is developing a more comprehensive partnering agreement. It governs the determination of obligations and costs to be borne by Caltrans and the utility companies on all freeways. Caltrans first used freeway master contracts in the 1950s when Street and Highway Code 707.5 was established and authorized Caltrans to enter into contracts with utility companies. Because of the large number of transportation projects at the time, there were many litigations on liability disputes. Officials believed the freeway master agreements would speed up the liability determination process. Consequently, nine agreements were executed and remained in place until 2003, when Caltrans entered into new freeway master contracts with eight utility owners. The major change was the uniformity of the terms and conditions among all contracts. Strengths of the new agreements included faster project delivery, a collaborative approach to working toward improvements in the project delivery process, and elimination of legal action on liability disputes. Weaknesses were that cost apportionments were still based on installation rights and that disagreements arose on interpretations of contract language. The new freeway master contracts also provided opportunities to reevaluate the effectiveness of the contract under current conditions and to renew and update the contracts in accordance with current budget allocations. Caltrans has faced an increase in the number of local projects involved with Caltrans right-of-way and challenges from local transportation agencies on the application of the freeway master contracts over the franchise area within the local project limits. A contract with Pacific Gas and Electric Company (PG&E) provided in this survey was signed by the PG&E executive vice president and chief of utility operations and the Caltrans chief of right-of-way and land surveys.

Caltrans has also partnered with the Inter-Utility Coordinating Council (IUCC). The IUCC is made up of representatives of major utility companies, represents the utility companies' interests and points of view in coordinating transportation projects, and improves communication and understanding among the stakeholders. The Caltrans–IUCC partnership has resulted in joint pilot programs with IUCC members, communication, conflict resolution, shared best practices, and discussions on ways for utilities to adapt to an increasing number of local projects. As a result of the IUCC partnering efforts, PG&E and Caltrans held high-level meetings to discuss broad challenges for both organizations. This resulted in a partnership to improve joint efforts and understanding between the two entities, which officials hope will result in a comprehensive partnering agreement similar to the Australian multilevel MOUs. Caltrans also established internal Utility Engineering Units in every district and a Utility Engineer Database. Utility Engineering Units are involved in the project process from conception to completion. Some districts have fully functioning units, while others are piloting this project delivery support activity. Caltrans is also developing a statewide utility database that will consolidate utility data, information in other databases, permit records, maintenance records, traffic records, and electrical records. Efforts are underway to establish requirements for all new utilities installed within the State's right-of-way.

Colorado

The Colorado DOT has a project-specific utility relocation agreement (PSURA) that it uses in conjunction with individual construction projects. A sample copy of a PSURA was provided in response to the survey. The PSURA is a master agreement that establishes a general framework for processing the utility work for utilities that need to be relocated or changed as necessitated by the project. Agreements are signed by the utility owners' authorized representatives and the highway agency's executive director.

Connecticut

The Connecticut DOT has a project-specific master agreement for the readjustment, relocation, and removal of utilities on highway projects. A sample copy was provided in response to the survey. These agreements are signed by utility owners' authorized representatives and the agency's chief engineer of the Bureau of Engineering and Highway Operations.

Delaware

The Delaware DOT has a public utility annual master franchise. An online link provided in response to the survey is in Appendix B. The master franchise eliminates the need to franchise each utility installation within State right-of-way. It does not change the conditions or the regulations utility companies must comply with for the privilege of occupying public right-of-way, nor does it alter the State's utility permitting process. It is intended solely to reduce paperwork and subsequent record storage. The only separate requirement on individual projects is the need for the utility companies to request permits to work on the highway right-of-way. Agreements are signed by the utility owners' authorized representatives and the agency's assistant director for engineering support.

Florida

The Florida DOT (FDOT) has numerous utility-related partnering agreements (joint use of utility facilities, utilities design-build, utility work at utility and FDOT expense combined, utility work by highway contractors at FDOT expense, etc.). A link to the FDOT agreements is in Appendix B. Agreements are signed by the utility owners' authorized representatives and the FDOT State utility engineer.

Louisiana

The Louisiana Department of Transportation and Development (LA DOTD) has standard project-specific agreements based primarily on whether the utility company will relocate on LA DOTD right-of-way or private right-of-way. LA DOTD also requires utility companies to submit relocation permits if they plan to work within DOTD right-of-way. If LA DOTD does not have any costs, it obtains drawings from the utility companies and sends an "approved drawings" letter. If the utility company plans to relocate on someone else's utility pole, LA DOTD sends a "waived drawing" letter. For utilities that are not in conflict, it sends a "no conflict" letter. Agreements are signed by the utility owner's authorized representatives and LA DOTD's utility relocation engineer.

Maine

A high-level MOU was executed in Maine in 1992. It was recently updated and the Maine DOT (MDOT) is implementing the changes. The 1992 MOU included most of the large utility companies in Maine at the time, but the updated MOU includes only the overhead utilities. A partnering session was planned to discuss the new MOU. MDOT provided both the 1992 and the latest version of the updated MOU in response to the survey. The update is believed to be the result of growing tension between overhead utilities and MDOT over utility relocation issues. Several factors have come into play on communications and scheduling, but the primary factor appears to be the amount of property being acquired for MDOT projects. The utility companies appear to believe they are being forced to the edge of the highway right-of-way and will not be able to easily obtain the additional property rights necessary to allow tree trimming outside the right-of-way. The 1992 MOU was developed by top-level MDOT officials, including a bureau director and deputy commissioner, and top management utility company officials. The MDOT commissioner signed the MOU. More than the wording of the MOU itself, which is somewhat broad, the discussions of concerns, organization, financing, and regulatory realities during the MOU negotiations have resulted in a renewed appreciation of each party's business needs and constraints. Also, those who signed the 1992 MOU have now retired or moved to new jobs and the updated MOU process has brought new players together. Ultimately, the new MOU sets a tone from the top that MDOT and businesses must cooperate to benefit society. A partnering session was planned to attempt to answer many of the remaining questions about changes resulting from the updated MOU.

Maryland

The Maryland State Highway Administration (MSHA) has several agreements with utility companies. A copy of a 1968 agreement with the Chesapeake and Potomac Telephone Company of Maryland was provided in response to the survey. MSHA's agreements are not detailed and multileveled. They do, however, establish requirements for adjusting and/or relocating existing utility facilities to facilitate construction of highway projects. The Chesapeake and Potomac agreement was signed by a company vice president and the MSHA chairman-director.

Minnesota

The Minnesota DOT has a project-specific master utility agreement (MUA) for design-build projects that covers all relocations required for that specific project. An online link, provided in response to the survey, is in Appendix B. These project-specific MUAs are signed by the utility owner's authorized representatives, the highway agency, and the contractor.

Missouri

The Missouri DOT has a master reimbursable utility agreement (MRUA). A copy of the MRUA was not provided. The MRUA is a generic agreement that addresses all future reimbursable utility relocations between a particular utility company and the DOT. Once an MRUA is executed, no other utility agreements are executed. All project-specific items such as type of agreement (lump sum or actual cost), estimated total cost, and cost allocation are addressed in a letter from the district utility engineer to the utility company. This agreement is typically executed by the director of program delivery or higher (chief engineer, chief operating officer, etc.).

New Jersey

The New Jersey DOT has a project-specific master agreement with all the major utility companies and various fiber optic companies for the adjustment and/or relocation of utilities in conjunction with highway projects. A copy of a master agreement and the change order for that agreement were provided in response to the survey. These agreements are signed by utility owners' authorized representatives and the agency's director of the Division of Project Management.

North Dakota

The North Dakota DOT (NDDOT) has a project-specific utility relocation agreement that it uses to pay utilities to relocate when NDDOT purchases additional right-of-way or obtains a construction easement that creates conflicts with the utilities and requires them to relocate. New State laws require NDDOT to coordinate and communicate with utility companies as early as possible when utility relocation is required. These agreements are signed by utility owners' authorized representatives and the NDDOT director of transportation.

Ohio

The Ohio DOT (ODOT) is working on a master agreement for utility coordination, estimating, and billing. It does not have the State law necessary to use the types of MOUs used in Australia, but officials believes this final report will provide them with the appropriate agreement between ODOT and the utility companies to focus both parties on the importance of efficient and effective utility relocation work. For now, the agreement deals with utility reimbursement when utilities are eligible for payment, but eventually ODOT will use a similar document for all projects. ODOT is developing an MOU because officials believe it is important to have an agreement in place to verify the encumbrance process for utility reimbursement. ODOT's utility section is using this opportunity to expand the document language to cover the utility relocation requirements (schedule, design, coordination, etc.) contained in its *Utilities Manual*. The current draft document was created by the agency's utility, financial, and legal offices. The utility companies have not been engaged in the process, but are expected to be in the future. ODOT is confident the utility companies will have no major problems with this effort because all it does is clearly define the utility relocation requirements in ODOT's *Utilities Manual*. Once the process is in place, ODOT will monitor its effectiveness and take it to the next level to cover all projects, regardless of whether the utility is in a reimbursable position.

Because the ODOT utilities engineer was involved in the 2008 international scan on right-of-way and utility practices, the agency's effort was influenced by the Australian processes, but the various levels of agreements are being handled by ODOT in a different way. Over the years, ODOT has maintained contact with middle and upper management of utility companies, which has been beneficial in resolving problems. ODOT's district utility coordinators never hesitate to involve the central office when they have utility relocation issues, which may occur any time during project initiation, design, or construction. A new ODOT effort involves holding meetings with central office leadership (director, assistant director, production deputy director), district deputy directors and their senior teams, and utility company leadership (presidents, vice presidents, chief engineers) to discuss mutual concerns and establish awareness of the importance of efficient utility coordination in project identification, design, and construction. Such a session was planned with the leadership of FirstEnergy Corporation, which has three subsidiaries, Cleveland Electric Illuminating Company, Ohio Edison Company, and Toledo Edison Company, that serve more than a third of Ohio residents.

Puerto Rico

The Puerto Rico Highway and Transportation Authority (PRHTA) has a master agreement for utility adjustments and relocations. Copies of master agreements with the two principal public utilities were provided: Puerto Rico Electric Power Authority (PREPA) and

Puerto Rico Aqueduct and Sewer Authority (PRASA). PRHTA also has a master agreement with a private utility: Puerto Rico Telephone. PRHTA, by law, must pay for PREPA and PRASA relocations. The master agreements were developed by midlevel PRHTA officials with high-level supervision and final decisionmaking, including the utilities office director of design and the deputy executive director for engineering. The same levels represented the utilities, with chief legal advisors helping with legal language and related matters and executive directors providing signatures.

Tennessee

For many years, the Tennessee DOT (TDOT) has had an agreement with the Tennessee Valley Authority (TVA) on utility installations and relocations for highway projects that affect TVA facilities. Because of TVA's quasigovernmental standing and the presence of TVA facilities in Tennessee, both agencies cooperatively agreed that such an arrangement would reduce coordination time and define responsibilities of both parties. Several years ago, TDOT revised the agreement to address issues that TVA indicated had evolved since the original agreement. At that time, legal counsel and agency coordinators were involved, and both agreed that the terms of the agreement should be expanded to include relocation so that individual relocation agreements for each project could be simplified. TVA is the only utility in Tennessee that writes its own relocation contract. Neither agency has pursued further modifications for this purpose, but TDOT intends to do so once resources are available. The agreement was signed by TVA's chief operations officer and TDOT's commissioner.

Texas

The Texas DOT (TxDOT) has had a utility MOU in place since 1998 that it uses with several utility companies. The MOU relies heavily on the TxDOT Utility Cooperative Management Process and a lower level subprocess that came about in 1995–1996 during a business process retooling effort.

It is a business model that describes such areas as parties, tasks, goals, objectives, and relationships. The purpose of the MOU is to establish relationships, clarify lines of communication, and outline the general procedure to accommodate public utility and common carrier use of public rights-of-way during construction of TxDOT transportation improvement projects. The MOU is intended to emphasize coordination and cooperation by all participants to both highway users and utility customers. The MOUs are used statewide, but they are nonbinding and voluntary. They are signed by TxDOT district engineers and utility executives. TxDOT is developing an updated MOU with a single utility company. Its primary purpose is to improve communication, cooperation, and coordination between utilities and TxDOT.

Virginia

The Virginia DOT (VDOT) has a master agreement for the adjustment and/or relocation of utility facilities on highway projects with 79 utility companies. A sample agreement with Dominion Virginia Power was provided. When VDOT authorizes a utility to relocate its facilities, the utility does not have to wait for VDOT to sign an agreement to begin work. VDOT simply references the conditions and date of the master agreement. These agreements contain procedures the utility company and VDOT must follow while constructing highway projects and sets terms and conditions under which the utility company will make necessary changes in its facilities and the State will reimburse the applicable costs incurred by such changes. The Dominion Virginia Power agreement was signed by the vice president for electric construction and the VDOT chief engineer.

West Virginia

The West Virginia DOT (WVDOT) has a master agreement that contains procedures the utility company and WVDOT must follow during the construction of highway projects and sets procedures and guidelines to be used when reimbursement of relocation costs is required by law. When

WVDOT requests that a utility company relocate facilities that conflict with highway construction, the utility company prepares detailed plans and estimates showing the work to be done to relocate its facilities, including temporary relocation, if necessary, for existing and proposed facilities. After WVDOT approves these plans and supporting data, the utility company may proceed within a reasonable time with relocation under the terms of the master agreement. These agreements are signed by the utility owner's authorized representative and WVDOT's deputy State highway engineer for development.

CHAPTER 4

Comparison of Australian MOUs and U.S. Partnering Agreements

Queensland and New South Wales in Australia have developed MOUs that meet their respective needs. Many States in the United States have also developed partnering agreements that meet their needs, but most admit that their agreements are not all that they could be and are interested in improving them.

The Australian MOU and the various partnering agreements in the United States are similar in some ways and different in others. One similarity and three significant differences are of primary importance to the United States:

- MOUs in Australia and partnering agreements in the United States are similar in that their primary purpose is to improve working conditions (coordination, cooperation, communication, and commitment) with utility companies. Highway officials in both countries recognize the importance of such relationships in advancing projects to completion.

   In New South Wales, the Streets Opening Conference, which started in 1909 as a focal point for discussing common transportation and utility issues, plays an important role in building highway and utility relationships. Membership includes utility owners, local government and road authorities, light rail operators, other government agencies, consultants, and other groups with an ongoing interest in utility issues. Its objectives include establishing recommended practices for providing utility services; fostering coordination; encouraging the use of agreed codes and practices for excavation, backfilling, and roadway reconstruction; and minimizing the impact of excavations. A major initiative it has undertaken is a model agreement for strategic alliances between utility and road authorities. This document defines provisions for notifications, work execution, restoration, and relocation of assets. It also outlines a policy and planning framework that includes coordination, performance standards, and dispute resolution. New South Wales' MOU is based on this model agreement.

   The United States has similar organizations in many states called Utilities Coordinating Committees (UCC). One particularly good one is the Florida Utilities Coordinating Committee. Established in 1932, this organization, which meets quarterly, is a confederation of public and private utilities, public works departments, consulting engineers, contractors, and State and local governmental agencies that work together to resolve problems and develop standards for coexistence in public rights-of-way. Recent major initiatives include significant review of and input on Florida DOT's *Utility Accommodation Manual* and development of a Utility Coordination Training Course for State and local departments of transportation, utility companies, and consultants.

- MOUs in Australia and partnering agreements in the United States are different in that the Australian MOUs were developed at a high organizational level, while the U.S. partnering agreements generally were developed at a middle level. The Australians firmly believe, and the Americans tend to agree, that policies filtering down from the highest levels in the organization are significantly more effective than those that may have questionable support from upper management. Some U.S. partnering agreements have been signed by high-level officials. This may have had some positive influence, but probably not as much as would have been provided if upper managers from both agencies had been involved in developing the partnering agreements.

- MOUs in Australia and partnering agreements in the United States differ on reimbursement for utility relocations. In Australia, states normally reimburse utility interests for relocating utility facilities because historically most utility owners have been government rather than private entities. In recent years, the Australian utility industry has been deregulated and a large percentage of utility interests are now in private hands, but the policy for reimbursing utility relocations has continued. In the United States, most States reimburse utilities for adjustments and relocations only if the utilities have prior rights (i.e., a property interest in the present location). A few States (Alaska, New Jersey, and Tennessee) have laws that give them the authority to pay 100 percent, and Montana pays 75 percent for all utility adjustments and relocations necessitated by highway construction. FHWA requires State DOTs to enter into agreements with utility companies only when reimbursement is involved, so most partnering agreements are primarily to establish prior rights and shares of the cost.

- MOUs in Australia and partnering agreements in the United States appear to have different requirements for noncompliance with the terms in the agreements. The Australians have generally included a shared-risk process in their MOUs in which participants share the risks and portions of the cost for not doing what the MOUs required. U.S. partnering agreements are often considered nonbinding and include termination clauses for noncompliance.

Other similarities and differences exist, but these three are the most important to any efforts in the United States to promote the Australian MOU concept. Most important, both State DOTs and utility companies must intend to cooperate, coordinate, and communicate. The term for this concept, CCC, was first used by TxDOT Right-of-Way Division Director John Campbell in the 1990s and has been widely used nationwide since then. All parties understand it to be the key to successful working relationships. CCC is essential and must be adopted by the highest level officials in both DOTs and utility companies. Utility-related laws, regulations, policies, and guidance have existed in the United States for almost 70 years. Even so, problems still exist, most related to a lack of CCC. These problems can be discussed, resolved, and documented in MOUs or similar partnering agreements.

CHAPTER 5

Findings and Recommendations

Findings

A survey was sent electronically to utility managers in 50 States, Puerto Rico, and the District of Columbia. The survey had two questions:

1. Does your State DOT have partnering agreements with utility companies that are similar to the Australian MOUs?

2. If your answer is yes, would you provide a sample copy and briefly explain why the agreement was initiated, what levels of the DOT and utility organizations were involved, and how the agreements have helped the relationship between the State and utility and improved the project development process?

There were 48 responses to the survey (46 States, the District of Columbia, and Puerto Rico). Only one State, Maine, responded that it had an MOU similar to the Australian MOU. Three States (California, Ohio, and Texas) are developing similar MOUs. Even though they did not have anything similar to the Australian MOUs, 20 States provided copies of master agreements, master contracts, MOUs, standard reimbursement agreements, and other partnering documents they use. Several States indicated that their partnering agreements work well for them. Most States, however, indicated that they would be interested in improving them and reviewing the step-by-step procedures and sample template developed as part of this study.

A followup question was e-mailed to those who responded to the original survey: If your answer was no would you be interested in being a pilot State to develop an MOU similar to the Australian MOU, or if your answer was yes would you be interested in upgrading your existing partnering agreement to something similar to the Australian MOU?

In response to this question, the following 10 State DOTs indicated interest in being pilot States:

- Arkansas
- California
- Illinois
- Missouri
- New Hampshire
- North Carolina
- Ohio
- South Carolina
- Tennessee
- Utah

The following six states indicated they might be interested in being pilot States:

- Hawaii
- Iowa
- Kentucky
- Massachusetts
- Michigan
- Vermont

Queensland and New South Wales have developed MOUs that meet their respective needs. Many States in the United States have also developed partnering agreements to meet their needs, but most concede that their agreements are not all that they could be and are interested in improving them. The Australian MOUs and the U.S. partnering agreements are similar in some ways and different in others. One similarity and three significant differences are important to the United States:

- They are similar in their primary purpose, which is to improve working conditions (coordination, cooperation, communication, and commitment) with utility companies.

- They are different in levels of development. The Australian MOUs were developed at a high organizational level, while U.S. partnering agreements generally were developed at lower organizational levels.

- They are different in reimbursement and access requirements. Utility companies are reimbursed 100 percent for relocations in Australia and some have unlimited access to the right-of-way, while in the United States DOTs have total control of the highway right-of-way and the majority reimburse utility companies for required relocations only if the utility companies have prior property rights.

- They are different in stated conditions for noncompliance with their partnering agreements. The Australians generally have included a shared-risk process, in which participants are responsible for meeting the terms of the MOUs or a shared portion of the costs of not doing so. In the United States, partnering agreements are often considered nonbinding and noncompliance provisions, if any, only involve termination clauses.

Other similarities and differences exist, but these are the most important to any efforts in the United States to promote the Australian MOU concept.

Recommendations

MOUs in the United States do not need to be as comprehensive or multilayered as those in Australia. This is because State and local DOTs in the United States manage the highway right-of-way and accommodate utilities in accordance with longstanding Federal, State, and local laws, regulations, policies, and guidance. Nonetheless, to advance highway projects to an earlier and less contentious completion, MOUs similar to the high-level MOUs used in Australia might be a considerable help. Implementation in the United States will most likely occur, however, only if DOT statewide utility managers champion the effort by doing the following:

- Consider the step-by-step procedures and the template in this report and the possibility of implementing some of the recommendations in a way that benefits their States.

- Begin the MOU process by discussing the MOU high-level concept with utility representatives at Utility Coordinating Committee meetings and by arranging meetings with high-level DOT officials to discuss the need for improved working relationships, as described in the step-by-step approach in Chapter 6.

- Continue the MOU process described in the step-by-step procedures and identify the primary issues, outcomes, implementation strategy, and benefits to be gained.

These MOUs will not replace the need in the United States for individual relocation agreements setting separate responsibilities for financing and accomplishing relocation work necessitated by highway construction. They may, however, reduce the need for much verbiage in the utility relocation agreements. This will be an issue for FHWA to address in the future based on the effectiveness of implemented MOUs in reducing paperwork and advancing projects to completion.

Partnering agreements in the United States are generally considered nonbinding documents describing agreements between parties, so it is imperative that these documents contain clauses describing actions to be taken in cases of noncompliance. It is recommended that in such cases, rather than terminate the MOU as is often done, parties should share any resulting costs (or benefits) associated with noncompliance and work together to amend the MOU, if necessary, to reflect conditions that both parties can accept.

CHAPTER 6

Step-by-Step Suggestions for Developing an MOU in the United States

A memorandum of understanding is defined as follows:

> A memorandum of understanding (MOU) is a document describing a bilateral or multilateral agreement between parties. It expresses a convergence of will between the parties, indicating an intended common line of action. It most often is used in cases where parties either do not imply a legal commitment or in situations where the parties cannot create a legally enforceable agreement. It is a more formal alternative to a gentlemen's agreement. (http://en.wikipedia.org/wiki/Memorandum_of_understanding)

Queensland and New South Wales in Australia have developed MOUs with major utility companies to facilitate the cooperation and coordination process and to optimize the relationship between transportation agency and utility interests. Many State DOTs in the United States are also looking for better ways to coordinate, cooperate, and communicate with utility companies. Memorandums of understanding similar to those used in Australia, but modified to reflect U.S. conditions, might be a way to do that. This chapter outlines steps for developing MOUs in the United States. These steps will not necessarily apply to conditions in all States, but are a useful starting point.

Step 1

It all begins with a champion, which in the United States will probably be the State DOT statewide utilities manager. This is the person who will recognize a need for better State and utility working relationships and believe that a MOU might be the best way for both parties to come together at a high organizational level to identify and resolve the major problems that keep them apart. This is something the manager might want to discuss at a Utilities Coordinating Committee meeting to get feedback from utility counterparts and with FHWA counterparts to get Federal involvement and buy-in early in the process.

Step 2

It may begin with a champion, but no matter how worthy the cause it will not go anywhere without support from the highest level officials in the organization. The statewide utilities manager must go through the chain of command to the chief engineer to discuss the need for better working relationships and why an MOU might be helpful and to request support to proceed with development of an MOU.

Step 3

Obtaining support from the chief engineer is only half the battle. The statewide utilities manager must now go to a counterpart at a major utility company to sell the MOU idea and encourage the counterpart to ask the utility's upper-level managers if they would meet with the DOT to discuss matters of mutual concern.

Step 4

Once the chief engineer has given the statewide utilities manager permission to develop an MOU and an interested utility company has been identified, the manager must schedule a meeting of high-level DOT and utility officials. For the DOT, those invited might include the commissioner, chief engineer, director of preconstruction, director of construction, director of right-of-way, and directors of other offices that deal with utilities. For the utility, those invited might include the chief executive officer, president, and vice presidents of offices involved in utility relocations necessitated by highway construction. A professional contract facilitator could be employed to help with the set up, agenda, planning, notification, and organization and, most important, to assist in brokering the MOU concept.

Step 5

Before the meeting of high-level DOT and utility officials, the statewide utilities manager must prepare an agenda of items to discuss. The most important item on the agenda would be a discussion of issues that impact working conditions at both agencies.

Step 6

At the meeting, the DOT commissioner will probably want the chief engineer or the office director responsible for utilities to act as facilitator. The facilitator should be well briefed by the statewide utilities manager. The outcome of the meeting should be an agreement to work together to identify, prioritize, and resolve major issues (i.e., issues rated highest in complexity and benefit to be gained by both parties) and to document these issues and agreed outcomes in an MOU. A steering committee of key representatives from each agency should also be appointed to oversee the work and a working group of technical experts should be appointed to do the actual work. The statewide utilities manager and his or her counterpart from the utility company should be prominent members of the working group.

Step 7

Now the real work begins. Top-level officials at the DOT and a major utility company have given their blessing, but now someone has to make it happen. Once again, it is the statewide utilities manager. The manager should convene a meeting of the working group to identify, prioritize, and resolve major issues. The working group members should elect a chair, a vice chair, and a recorder at the first meeting. The chair will facilitate the first and all subsequent meetings of the group and coordinate all activities with the steering committee. If the chair is not the statewide utilities manager, the manager must nonetheless be the force behind the working group to assure that it keeps moving forward and coordinates effectively with the steering committee.

Step 8

Once the working group has identified and prioritized major issues impeding good working relationships and these issues have been discussed and resolved through agreed outcomes by all members of the working group and the steering committee, another meeting of high-level officials should be convened. The statewide utilities manager should schedule the meeting and the steering committee should present the issues and outcomes for upper-management consideration.

Step 9

The major issues of contention between the DOT and the utility have been identified, prioritized, discussed, and resolved by midlevel officials from both agencies. Now it is time for the leadership to take a look at them. This may take several meetings. Sometimes things look different at the top than from the middle or bottom. Hopefully, the leaders of both organizations will approve the steering committee's recommendations and authorize it to document the approval in an MOU that identifies the primary issues, agreed outcomes, implementation strategy, and benefits to be gained. The MOU will be signed by the commissioner of the DOT and the chief executive officer of the utility.

Step 10

Once the MOU has been signed, it must be widely promoted to all DOT and utility personnel who deal with the adjustment or relocation of utility facilities and must be implemented in the working operations of both agencies. The statewide utility manager and the utility counterpart must monitor work being performed to assure that provisions of the MOU are carried out as intended.

Step 11

One year after the MOU goes into effect, the statewide utility manager and utility counterpart should evaluate the results to see if any improvements have been made in working relationships and the advancement of construction projects. These results, positive or negative, should be reported through the steering committee to top-level officials of the DOT and utility organizations.

Step 12

If results have been positive, efforts should be made to establish similar MOUs with other utility companies.

Sample template that can be used to develop a Department of Transportation–Utility MOU

Memorandum of Understanding Between State DOT and Utility Company

SUBJECT:
Highway-Utility Memorandum of Understanding

Purpose

The purpose of this memorandum of understanding is to facilitate the cooperation, coordination, and communication process and to optimize the relationship between highway agency and utility company interests.

References

References that are directly related to this MOU are as follows:

- *Streamlining and Integrating Right-of-Way and Utility Processes with Planning, Environmental, and Design Processes in Australia and Canada* (FHWA-PL-09-011), U.S. Department of Transportation, Federal Highway Administration, June 2009.

- *Program Guide: Utility Relocation and Accommodation on Federal-Aid Highway Projects* (FHWA-IF-03-014), U.S. Department of Transportation, Federal Highway Administration, January 2003.

- *A Policy on the Accommodation of Utilities Within Freeway Right-of-Way*, American Association of State Highway and Transportation Officials, October 2005.

- *A Guide for Accommodating Utilities Within Highway Right-of-Way*, American Association of State Highway and Transportation Officials, October 2005.

- *Right-of-Way and Utilities Guidelines and Best Practices*, Strategic Plan 4-4, American Association of State Highway and Transportation Officials, Standing Committee on Highways, January 2004.

Problem

Historically, it has been in the public interest for utilities to use and occupy the right-of-way of public roads and streets. This is especially the case on local roads and streets that primarily provide a land service function to abutting residents, as well as on conventional highways that serve a combination of local, State, and regional traffic needs. Over many years it has been the most feasible, economic, and reliable solution for transporting people, goods, and public service commodities (water, electricity, communications, gas, oil, etc.), all of which are vital to the general welfare, safety, health, and well-being of the American people.

No matter how advantageous it has been for highways and utilities to share the right-of-way, many issues have arisen among the parties. These issues have been addressed by the Federal Highway Administration and the American Association of State Highway and Transportation Officials in regulations, policies, and guidelines. Even so, the basic issue—working relationships—remains troublesome for both highway and utility agencies and often hampers the operations of both organizations, resulting in time-consuming and costly project delays.

[Provide additional information on highway agency and utility company background and working relationships, both positive and negative.]

Scope

This MOU applies to all statewide highway projects that necessitate the adjustment and/or relocation of utility facilities.

Issues, Agreements, Support and Resource Needs, Implementation, and Benefits

[Document the primary issues discussed, agreed outcomes, support and resource needs, implementation procedures, and benefits to be gained.]

Monitoring

One year after this MOU goes into effect, the highway agency utilities manager in conjunction with the utility company counterpart will evaluate the results to see if any improvements have been made in working relationships and the advancement of construction projects. These results, positive or negative, are to be reported to signers of this document with a recommendation that they be discussed and resolved.

Conflict Resolution

This MOU may be amended or supplemented by mutual agreement between the parties and may be terminated by either party through a written notice to the other party. All parties are responsible for meeting the terms of this MOU. If they do not, _____ percent of the cost of not doing what they were required to do becomes their responsibility. Conversely, if project savings should result, those savings are to be shared on a similar percentage basis. A conflict resolution process is attached in matrix form (see Attachment A). [Note: A sample conflict resolution matrix is in Appendix D.] The intent of this process is to resolve each contract claim or dispute within a reasonable amount of time and at the organizational level closest to the source of the problem. Uniformity is important in notification and documentation and in providing the consultant adequate opportunity to participate in resolving the issue. This process is needed to insure that both parties' rights are protected if an issue is not resolved and continues in the review and appeals process.

Effective Date

This memorandum of understanding has been approved and signed by the parties below and becomes effective on

_____.
(date)

_____
Commissioner, State DOT

_____
CEO, Utility Company

References and Resources

A Guide for Accommodating Utilities Within Highway Right-of-Way, American Association of State Highway and Transportation Officials, October 2005.

A Policy on the Accommodation of Utilities Within Freeway Right-of-Way, American Association of State Highway and Transportation Officials, October 2005.

Avoiding Utility Relocations: Making the Effort Work (FHWA-IF-02-048), U.S. Department of Transportation, Federal Highway Administration, Office of Program Administration, July 2002.

Kirk, James E., *Utility Relocation and Accommodation: A History of Federal Policy Under the Federal-Aid Highway Program*, U.S. Department of Transportation, Federal Highway Administration, June 1980.

Program Guide: Utility Relocation and Accommodation on Federal-Aid Highway Projects (FHWA-IF-03-014), U.S. Department of Transportation, Federal Highway Administration, January 2003.

Right-of-Way and Utilities Guidelines and Best Practices, Strategic Plan 4-4, American Association of State Highway and Transportation Officials, Standing Committee on Highways, January 2004.

Standard Guideline for the Collection and Depiction of Existing Subsurface Utility Data (CI/ASCE 38-02), American Society of Civil Engineers, 2003.

Streamlining and Integrating Right-of-Way and Utility Processes With Planning, Environmental, and Design Processes in Australia and Canada (FHWA-PL-09-011), U.S. Department of Transportation, Federal Highway Administration, June 2009.

APPENDIX A

Summary of Australian Memorandums of Understanding

Queensland		New South Wales
Energex	**Telstra**	**Sydney Water**
Construction and Maintenance ■ Sequencing ■ Relocation timing ■ Future costs ■ Location details Corridor Planning, Access, Infrastructure Design ■ Strategic planning ■ Securing alignments ■ Future costs ■ Underground versus overhead	Asset Protection (AP) ■ AP procedures ■ AP timing ■ Telecommunications alignment ■ AP costs Access to Roads ■ Land access notification contract ■ AP contract ■ Damage minimization	Framework ■ Cost distribution ■ Information sharing ■ Strategic planning ■ Project management ■ Dispute resolution ■ Typical situations ■ Additional information (e.g., agreement and cost distribution)
An MOU was developed identifying the primary priorities group, with targeted agreed outcomes, the benefits to be gained, and the degree of complexity to achieve these goals.		An MOU was developed based on the New South Wales Streets Opening Conference model.
All agreed outcomes were signed by either the chief executive officers of the respective organizations or by the steering committee and implemented in the working operations.		All agreed outcomes were signed by the chief executive officers of the respective organizations and implemented in the working operations.

APPENDIX B

Summary of Responses to State Department of Transportation Survey

Question 1: Does your State department of transportation (DOT) have a partnering agreement with a utility company that is similar to the Australian memorandum of understanding (MOU)?

Question 2: If your answer is yes, would you provide a sample copy?

Question 3: If your answer is no, would you be interested in being a pilot State to develop an MOU similar to the Australian MOU, or if your answer is yes, would you be interested in upgrading your existing partnering agreement to something similar to the Australian MOU?

State DOT	Contact	Multilevel MOU Response	Pilot Response
Alabama	Robert Lee	**NO** We don't have an MOU with the utility companies. We execute project-specific agreements on each project where utilities are in conflict. For the reimbursable work, we need an estimate and the work is often done by low-bid contract. The agreement is also helpful if the utility wants to include betterment. (I don't know how that would work with an MOU.) For the nonreimbursable work, we have an agreement that doesn't take very long to get approved, lessening any time advantage of an MOU.	**NO** To adopt MOUs would require a process of review by the utility companies and some negotiation between our legal staff and that of the utilities. I would have to be sold on the benefits of the MOUs to make the effort to implement them. I would also have to sell upper management on the idea. Unless there are obvious and easily understood advantages I don't know if my State would be a good candidate.
Alaska	Ken Morton	**NO** **Master Agreements** The Alaska DOT does not have high- or midlevel MOUs with utility companies similar to what was found in Australia. We regularly do put together contract-level agreements to address reimbursable relocations. Upper management of the ADOT and the utility companies do not generally interact. Preconstruction coordination occurs at a lower level. The highest type of MOU (except utility agreements for reimbursable relocations) we have between the ADOT and a utility company occurs with design-build projects.	

State DOT	Contact	Multilevel MOU Response	Pilot Response
Arizona	Mona Aglan	**NO** **Master Agreement** We have a standardized agreement with Salt River Project and a standardized agreement that we use with the other utility companies. Attached is the standardized agreement with SRP.	
Arkansas	Perry Johnston	**NO** Arkansas does not have any kind of a partnering or master agreement similar to the high- or midlevel Australian MOU.	**YES** We are beginning the process of updating our utilities manual and accommodation policies. As part of the updating process, we have already had some preliminary conversations with some of the large utility companies about the concept of a multilevel MOU. We would be very interested in being a pilot State, and we are willing to offer any assistance we can to your development efforts. Thank you for the opportunity to be considered as a pilot State.
California	Lorrie Wilson	**NO** **Freeway Master Agreements** **Partnering Agreement (Underway)** I have attached a copy of the freeway master contract that addresses cost apportionment on freeway projects. The freeway master contract is in the midst of a major change due to the local agency's protest in paying any portion of utility relocation costs in franchise areas outside the right-of-way. We began working on a partnering agreement with Pacific Gas and Electric Company, but haven't moved along enough to be able to share anything at this point. See Paul Phan's presentation for information about both freeway master agreements and the new partnering agreement being developed.	
Colorado	Dahir Egal	**NO** **Utility Relocation Agreement** This is the closest CDOT has for utility relocation agreements.	
Connecticut	Sohrab Afrazi	**NO** **Master Agreement** The answer to your question is yes (see attached sample copy of our existing agreement). Currently, the Connecticut Department of Transportation is in the process of updating its master utility agreement.	**NO** Not interested at the present time.
Delaware	Fran Hahn	**NO** **Master Agreement** Here is the link: www.deldot.gov/information/business/public_utility/pdf/master_franchise_agreement_2009.pdf	

(continued)

State DOT	Contact	Multilevel MOU Response	Pilot Response
District of Columbia	Ardeshir Nafici	**NO** DDOT does not have a master agreement with utility companies.	
Florida	Gordon Wheeler	**NO** **Master Agreements** FDOT uses master agreements to establish certain guidelines to operate within so that approval by a board of directors or a city council is not necessary before utility work planning can begin. The utility work schedule is also used to plan the progress of the utility work. I don't know if the agreements qualify as a high-level MOU, but it seems to work for us.	**NO**
Georgia	Jeff Baker	**NO**	
Hawaii	Dean Yogi	**NO** We do not have a master agreement yet with utilities, although we keep trying. If you have any good ideas on how to accomplish this we would appreciate any help or suggestions. If we can get a copy of the MOU when it is ready, we would also appreciate a look. Legislation is a challenge. Agreeing to constantly changing language is also a challenge. A master-type agreement poses past and future effects. Thanks for your inquiry and including our State.	**MAYBE** If we can get a copy of the MOU when it is ready, we would also appreciate a look.
Idaho	Jack Masitis	**NO** Idaho does not have an MOU with utility companies	
Illinois	Cheryl Cathey	**NO** Illinois does not have a high-level master agreement with utilities. We are trying with railroads with mixed results.	**YES** We would be very interested in trying to get upper management to buy in. They are aware and are always looking for ways to minimize the delays.
Indiana	Matt Thomas	**NO** Indiana does not have such an agreement.	**NO** I am not interested. Thank you for asking though.
Iowa	Gerry Ambroson	**NO** Iowa has no MOU or partnering/master utility relocation agreements.	**MAYBE** We might be interested. Please provide a copy of the final guidelines and sample MOU when they are available.
Kansas	Mitch Sothers	**NO** Kansas DOT does not use a partnering or master agreement with utilities. We use the project-specific contract-level agreements.	

State DOT	Contact	Multilevel MOU Response	Pilot Response
Kentucky	Jennifer McCleve	**NO** The KYTC does not have any kind of partnering or master agreement with utilities like the high- or midlevel Australian MOUs. Our agreements are purely project specific in nature. These agreements reference our State laws governing the utility relocation process and our construction standards of interest. These agreements are drafted by regional representatives and executed by our Central Office. I see some value to master agreements and would be interested in receiving more information on Australia's MOU process.	**MAYBE** I have recently taken this position and have decided to take this first 6 to 9 months to fully assess the process we have at hand. I don't want to change or add processes until I fully inventory what we do now. So, I may not be the best champion at the time. However, I would certainly request that you provide me the material necessary to determine if we should proceed in this direction thereafter. It is certainly of interest to me.
Louisiana	JoAnn Kurts	**NO** **Standard Agreements** I am unaware of LA DOTD having an MOU. We have standard agreements primarily based on whether the utility company will be relocating into DOTD right-of-way or private right-of-way.	
Maine	Bill Pulver	**YES** **Memorandum of Understanding** I would say we have had both the high-level MOU and the contract-level agreement for some time, but not the midlevel. A high-level MOU was executed in 1992. I believe it is the first one we did. We have recently updated the high-level MOU, but are just now going through the process to implement the changes. It should be noted the 1992 MOU included most large utility companies in Maine at the time, but the updated MOU only includes the overhead utilities. We are planning a partnering session to discuss the new high-level MOU in the next couple of months. I have attached both the 1992 and what I have as the latest version of the updated high-level MOU. The project agreement in Appendix A of the January 5, 2009, MOU should not be considered a final version.	
Maryland	Nelson Smith	**NO** **Master Agreement** Maryland has several master agreements with utility companies. I've attached one that we have with the phone company. I'm not sure, however, how useful they will be to you. Based on your e-mail regarding this subject on the Australian master agreements, our agreements are not that detailed and not multileveled.	

(continued)

State DOT	Contact	Multilevel MOU Response	Pilot Response
Massachusetts	Guy Rezendes	**NO** Unfortunately, I don't have any examples of an MOU that you're looking for. We are working on a few new things insofar as earlier coordination with the utilities and consultants, etc., but nothing like this.	**MAYBE** I'll give you the typical "maybe" response. We have about three pilot project things going on just with utilities, and the other sections are involved as well. But keep me posted. I'll be able to give you a more descriptive answer later, or once I have these other pilots underway.
Michigan	Mark Dionise	**NO** Michigan DOT does not have agreements or MOUs similar to those used in Australia states. We only have the required agreements needed for reimbursable utility relocations. These agreements do not include any language similar to MOUs used in Australia.	**MAYBE** I would be very interested in pursuing this further. Recently, we have developed a Michigan Utility Coordination Conference (MIUCC) where we discuss issues like this with utility company representatives, MDOT utility coordinators, contractors, and MDOT consultants. This would be a good topic to bring to that group.
Minnesota	Marilyn Remer	**NO** **Master Agreements for D-B** We don't have master utility agreements, except for design-build projects on which we have master utility agreements (MUAs) that cover all relocations required for that specific project. Those MUAs can be accessed from our utility Web site as part of the design-build supplement at www.dot.state.mn.us/utility/files/pdf/policy/design-build-supplement-web.pdf.	
Mississippi	No response and no apparent Web site information		
Missouri	Jim Zeiger	**NO** To my knowledge, MoDOT has no high-level MOUs with any utility companies. We do have the master reimbursable utility agreement (MRUA) typically executed by the director of program delivery or higher (chief engineer, chief operating officer, etc.). I'm not sure if this is what you're referring to, but hope it helps.	**YES** Actually, it might be something to look at. We've made numerous improvements in how we deal with utilities and the flexibility we allow in using industry state-of-the-art materials, among other things. Something like this might provide the documented commitment from both sides. Yes, I'd like for Missouri to be considered. Naturally it would depend on MoDOT leadership's willingness to be involved.
Montana	Walt Scott	**NO** Sorry, but I cannot help you out with this one. We are working on an MOU with BNSF Railroad, but we have not completed it yet.	**NO** Upper management doesn't want to enter into master agreements for two reasons: (1) we are already overburdened with work, and (2) the present working relationship with utilities is very good and they don't want to tamper with it.

State DOT	Contact	Multilevel MOU Response	Pilot Response
Nebraska	Mark Ottemann	NO	
Nevada	Paul Saucedo	NO I do not know of any master agreements that we currently have with any utility company.	
New Hampshire	Chuck Schmidt	NO The New Hampshire DOT doesn't have any master agreements.	YES The New Hampshire DOT would be interested in being a pilot State.
New Jersey	Frank Pinto	NO **Master Agreements** NJDOT does have master agreements with all the major utility companies and various fiber optic companies. Attached is a copy of a master agreement and the change order for that master agreement.	
New Mexico	No response and no apparent Web site information		
New York	Mike Mariotti	NO Thanks for the inquiry. NYSDOT does not have any such master agreements with utilities. Our utility agreements are limited to project-specific contract-level agreements detailing contract-specific provisions.	
North Carolina	Robert Memory	NO We have no such agreement. We've discussed formatting an MOU agreement for the past several years. Unfortunately no action has been taken.	YES!!! NCDOT would like to participate as one of the pilot States.
North Dakota	Monte Dockter	NO NDDOT does not have any agreement like what is described in place at this time. The only agreement we have is when we pay the utility to relocate when we purchase additional right-of-way or get a construction easement that creates conflicts with the utility and requires it to relocate.	I will discuss with upper management and get back to you.

(continued)

State DOT	Contact	Multilevel MOU Response	Pilot Response
Ohio	Ray Lorello	**NO** **Partnering Agreement (underway)** ODOT is currently working on a master agreement for utility coordination, estimating, and billing. It doesn't have the State law necessary to do the types of MOUs used in Australia, but believes this document will provide the department with the appropriate agreement between the department and the utility to bring focus for both parties on the importance of efficient and effective utility relocation work. The agreement, for now, deals with utility reimbursement when utilities are eligible for payment, but we will eventually use a similar document for all projects, taking one step at a time.	**YES**
Oklahoma	Kurt Harms	**NO** ODOT does not have any master agreements in place with any utility owners. We work through the standard process for each relocation: field meeting, proposal, utility relocation agreement, work order, relocation.	
Oregon	Heather Howe	**NO** We do not have a master agreement with a major utility company.	
Pennsylvania	Larry Ditty	**NO** Pennsylvania DOT does not have any kind of a partnering or master agreement with utilities similar to the high- or midlevel Australian MOU.	**NO** As much as I would like to be one of the pilot States, I cannot commit at this time because of our current workload and being short-staffed.
Puerto Rico	Pedro Alvarado	**NO** **Master Agreements** Included are master agreements with the two principal public utilities, Puerto Rico Electric Power Authority (PREPA, whole electric system on the island) and Puerto Rico Aqueduct and Sewer Authority (PRASA, whole water and sanitary system on the island), and with Puerto Rico Telephone (main telecommunications company, private). We have to pay, by law, PREPA and PRASA relocation. These master agreements were worked by midlevel officials with high-level supervision and final decision-making. In our case, I did it as the Utilities Office director of design area (career position) and the deputy executive director for engineering of PR Highway Authority at that moment. Basically the same corresponding levels represented the utilities, with proper chief legal advisors helping with legal language and related matters, up to formal signature by the executive directors.	

State DOT	Contact	Multilevel MOU Response	Pilot Response
Rhode Island	No response and no apparent Web site information		
South Carolina	Mark Attaway	**NO** We do not have any kind of partnering agreement or MOU with the utility companies.	**YES** We would like to promote a pilot with upper management.
South Dakota	Dave Hausmann	**NO** SDDOT doesn't have any agreements other than the ones used for relocation reimbursement. We have discussed creating an MOU high- or midlevel type agreement that the SDDOT and involved utility company would use during my Advanced Utility Coordination Process, but haven't completed that task yet.	
Tennessee	Joe Shaw	**NO** **Master Agreement** TDOT has for many years had in place an agreement with TVA (Tennessee Valley Authority) regarding utility installations and relocations for highway projects that affect TVA facilities. Due to the nature of TVA's quasigovernmental standing and the presence of TVA facilities in Tennessee, cooperatively both agencies agreed that such an arrangement would reduce the coordination time and define responsibilities of both parties.	**YES** I am open to any improvements to the process. In the local FHWA office, Charlie O'Neill is over the right-of-way and utility program and would have to be open to a new process as well. I think TDOT would be willing to participate. We have some major utility groups in Tennessee that I would be interested in approaching, including the Memphis Light, Gas and Water, Knoxville Utility Board, and Nashville Electric Service, and of course I'd like to improve the TVA arrangement.

(continued)

Appendix B: Summary of Responses to State Department of Transportation Survey

State DOT	Contact	Multilevel MOU Response	Pilot Response
Texas	Jesse Cooper Randy Anderson	**NO** **Memorandum of Understanding Partnering Agreement (Underway)** The Texas DOT has a utility MOU that has been in place since 1998 and used with several utility companies. The MOU relies heavily on the TxDOT Utility Cooperative Management Process, also known as "The Process," and a concurrently running but lower-level subprocess that came about in 1995–1996 during a business process retooling effort. It is basically a business model that describes parties, tasks, goals, objectives, relationships, etc. The purpose of the MOU is to establish relationships, clarify the lines of communication, and outline the general procedure to accommodate public utility and common carrier use of public rights-of-way during construction of TxDOT transportation improvement projects. The MOU is intended to emphasize coordination and cooperation by all participants with the anticipated result of mutual benefit to both highway users and utility customers. The MOUs are used statewide. They are, however, nonbinding and voluntary. They are signed by TxDOT district engineers and utility executives. TxDOT is in the process of developing an updated MOU with a single utility company. Its primary purpose is to improve communication, cooperation, and coordination between utilities and TxDOT.	**NO**
Utah	Justin Sceili	**NO** No, but we are currently working on establishing them.	**YES** I believe we would be a great State to try it. We have already greased the skids and talked to our bigger utility companies and they are on board. I would love to participate any way possible.
Vermont	Craig Keller	**NO** Vermont is currently doing only project-specific utility agreements	**MAYBE** Things are very busy here with downsizing (budgetary constraints) and stimulus funding, but yes, I have interest in such a process.
Virginia	Greg Wroniewicz	**NO** **Master Agreements** We have master agreements with 79 utilities like the one attached with Dominion Virginia Power. When we authorize the utility to relocate its facilities, we do not have to wait for it to sign an agreement before it begins work. We simply reference the conditions and date of the master agreement. This gets the process moving faster.	

State DOT	Contact	Multilevel MOU Response	Pilot Response
Washington	Ahmer Nizam	**NO** The only utility-related master agreement that WSDOT uses is specifically for subsurface utility engineering (SUE) contractors, but not with the utility companies themselves. For SUE work, a contractor enters into a task agreement under the master agreement, and is then eligible to be selected as a SUE contractor for WSDOT projects. This appears to be different than what you are looking for, but we can provide more information if you are interested.	
West Virginia	Sarah Daniel	**NO** **Master Agreements** Attached is our master agreement template. If you need a copy of an actual master agreement, let me know.	
Wisconsin	No response	**NO** WisDOT Web site contains an Agreement for Payment for the relocation or replacement of certain utility facilities on publicly held lands required by the construction of transportation improvement projects.	
Wyoming	Ken Keel	**NO** WYDOT does not use any master agreements.	

APPENDIX C

Summary of U.S. Partnering Agreements

State	Type of Agreement	Parties to Agreement	Discussion of Agreement	Agreement Signers
Alaska	Project-specific memorandum of understanding (MOU)	Alaska DOT and Enstar Natural Gas Company	This MOU establishes the basis for reimbursement through the department for preliminary engineering activities by the utility for a replacement-in-kind relocation in accordance with provisions of Alaska Statute (AS) 19.25.020 and Alaska Administrative Code (AAC) Title 17, Chapter 15, Title 3 Utility Relocation and Adjustment.	ADOT utility section chief and utility representatives
Arizona	Project-specific master utility agreement	Arizona DOT and Salt River Project	The purpose of this agreement is to relocate, adjust, or rearrange certain existing facilities belonging to SRP to facilitate construction of the project by the Arizona Department of Transportation. SRP's facilities to be relocated have prior rights.	ADOT engineer-manager of utility section and SRP department manager
California	Freeway master contract (in place) Partnering agreement (underway)	California DOT and Pacific Gas and Electric	The freeway master contract, in accordance with the provisions of Section 707.5 of the Streets and Highways Code, governs the determination of the obligations and costs to be borne by the department and the utility on all freeways for utility work described in the contract.	PG&E executive vice president for utility operations and California DOT chief, Division of Right-of-Way and Land Surveys
Colorado	Project-specific utility relocation agreement (PSURA)	Blank copy	This PSURA is a master agreement that establishes a general framework for processing the utility work for utilities owned by the owner that need to be relocated or changed for the project.	Utility owner's authorized representative and CDOT executive director
Connecticut	Project-specific master agreement	Blank copy	For readjustment, relocation, and/or removal of utilities on highway projects	Utility owner's authorized representative and CDOT's chief engineer, Bureau of Engineering and Highway Operations
Delaware	Public utility annual master franchise	Blank copy	The DelDOT public utility annual master franchise eliminates the need to franchise each utility installation within State right-of-way. It does not change the conditions or the regulations with which utility companies must comply for the privilege of occupying public right-of-way, nor does it alter the State's utility permitting process. It is intended solely to reduce paperwork and subsequent record storage.	Utility owner's authorized representative and DelDOT assistant director for engineering support

State	Type of Agreement	Parties to Agreement	Discussion of Agreement	Agreement Signers
Florida	Numerous agreements	Blank copy	http://formserver.dot.state.fl.us/capture/listings/FormListing.aspx?office=UTILITIES	Utility owner's authorized representative and FDOT representative
Louisiana	Project-specific utility relocation agreement	Blank copy	LA DOTD has standard agreements primarily based on whether the utility company will relocate into DOTD or private right-of-way. In addition, LA DOTD requires a utility company to fill out a relocation permit if it will be within DOTD right-of-way. If LA DOTD does not have any costs, it gets drawings from the utility companies and sends out an "approved drawings" letter. If the utility company plans to relocate on someone else's utility pole, LA DOTD sends out a "waived drawing" letter. For utilities that are not in conflict, there is a "no conflict" letter.	Utility owner's authorized representative and LA DOTD's utility relocation engineer
Maine	MOU	Blank copy	A high-level MOU was executed in Maine in 1992. It was recently updated and MDOT is implementing the changes. The 1992 MOU included most of the large utility companies in Maine at the time, but the updated MOU includes only the overhead utilities. A partnering session was planned to discuss the new high-level MOU. More than the wording of the MOU itself (which is somewhat broad and general), the discussions of concerns and organizational, financial, and regulatory realities during the MOU negotiations bring a renewed appreciation of each party's business needs and constraints. Also, every person who signed the 1992 MOU has retired or moved to new jobs. The updated MOU process brought new players together. Ultimately, the new MOU sets a tone from the top that MDOT and businesses need to cooperate for the benefit of society. The partnering session was planned to attempt to answer many of the remaining questions about changes in the updated MOU.	For DOT, the 1992 MOU was developed by top-level officials: a bureau director and deputy commissioner For utilities, top management was involved and the MDOT commissioner signed the MOU
Maryland	Agreement	MSHA and C&P of Maryland	The Maryland State Highway Administration has several agreements with utility companies. A copy of a 1968 agreement with the Chesapeake and Potomac Telephone Company of Maryland was provided in response to the survey. MSHA's agreements are not detailed and multileveled. They do, however, establish requirements for adjusting and/or relocating existing utility facilities to facilitate construction of highway projects.	C&P vice president and MSHA chairman-director

(continued)

Appendix C: Summary of U.S. Partnering Agreements

State	Type of Agreement	Parties to Agreement	Discussion of Agreement	Agreement Signers
Minnesota	Project-specific master utility agreement (MUA) for D-B projects	Blank copy	The Minnesota DOT has a project-specific MUA for design-build projects that covers all relocations required for that specific project. MUAs can be accessed at the Mn/DOT utility Web site in the design-build supplement at www.dot.state.mn.us/utility/files/pdf/policy/design-build-supplement-web.pdf.	Utility owner's authorized representative, Mn/DOT, and contractor
Missouri	Master reimbursable utility agreement (MRUA)	None	The MRUA is a generic agreement that addresses all future reimbursable utility relocations between a particular utility company and the DOT. Once an MRUA is executed, no other utility agreements are executed. All project-specific items, such as type of agreement (lump sum or actual cost), estimated total cost, and cost allocation, are addressed in a letter from the district utility engineer to the utility company.	Utility owner's authorized representative and MoDOT director of program delivery or higher (chief engineer, chief operating officer, etc.)
New Jersey	Project-specific master agreement	Blank copy	NJDOT has a master agreement with all the major utility companies and various fiber optic companies.	Utility owner's authorized representative and NJDOT director, Division of Project Management
North Dakota	Project-specific utility relocation agreement	Blank copy	The North Dakota DOT has a project-specific utility relocation agreement that it uses to pay the utility when NDDOT purchases additional right-of-way or obtains a construction easement that creates conflicts with the utilities and requires them to relocate.	Utility owner's authorized representative and NDDOT director of transportation
Ohio	MOU (underway)	None	The Ohio DOT is working on a master agreement for utility coordination, estimating, and billing. It does not have the State law necessary to use the types of MOUs used in Australia, but believes this document will provide it with the appropriate agreement between the department and the utility to focus both parties on the importance of efficient and effective utility relocation work. The agreement deals with utility reimbursement when utilities are eligible for payment, but the agency will eventually use a similar document for all projects.	Not yet determined

State	Type of Agreement	Parties to Agreement	Discussion of Agreement	Agreement Signers
Puerto Rico	Master agreement	PRHTA and PREPA PRHTA and PRASA	The Puerto Rico Highway and Transportation Authority (PRHTA) has a master agreement it uses for utility adjustments and relocations. Copies of master agreements with the two principal public utilities were provided: Puerto Rico Electric Power Authority (PREPA, (whole electric system on the island) and Puerto Rico Aqueduct and Sewer Authority (PRASA, whole water and sanitary system on the island). PRHTA also has a master agreement with a private utility, Puerto Rico Telephone (PRT, main telecommunications company). PRHTA, by law, must pay for PREPA and PRASA relocations. The master agreements were developed by midlevel officials with high-level supervision and final decisionmaking, including the Utilities Office director of design and the deputy executive director for engineering of PRHTA. Corresponding levels represented the utilities, with chief legal advisors helping with legal language and related matters, and the executive directors provided signatures.	Executive directors
Tennessee	Master agreement	Tennessee DOT and Tennessee Valley Authority (TVA)	The master agreement sets agreed-on provisions for carrying out guiding principles in a Federal Highway Administration–TVA MOU dated August 27, 1970. The master agreement contains procedures that TVA and Tennessee DOT have agreed to follow during the construction of highway projects.	TVA chief operations officer and TDOT commissioner
Texas	Utility MOU (in place) Updated MOU (in progress)	Various utilities	The Texas DOT has had a utility MOU in place since 1998 and has used it with several utility companies. The MOU relies heavily on the TxDOT Utility Cooperative Management Process and a lower-level subprocess that came about in 1995–1996 during a business process retooling effort. It is basically a business model that describes parties, tasks, goals, objectives, relationships, etc. The purpose of the MOU is to establish relationships, clarify the lines of communication, and outline the general procedure to accommodate public utility and common carrier use of public rights-of-way during construction of TxDOT transportation improvement projects. The MOU is intended to emphasize coordination and cooperation by all participants with the anticipated result of mutual benefit to both highway users and utility customers. The MOUs are used statewide, but are nonbinding and voluntary. TxDOT is developing an updated MOU with a single utility company. Its primary purpose is to improve communication, cooperation, and coordination between utilities and TxDOT.	TxDOT district engineers and utility executives

(continued)

State	Type of Agreement	Parties to Agreement	Discussion of Agreement	Agreement Signers
Virginia	Master agreement	Virginia DOT and Dominion Virginia Power Company	The Virginia DOT has a master agreement for the adjustment and/or relocation of utility facilities on highway projects with 79 utility companies. These agreements contain procedures the utility and the department must follow during the construction of highway projects and sets terms and conditions under which the utility will make necessary changes in its facilities and the State will reimburse the utility the applicable costs incurred by such changes.	Vice president of electric construction and VDOT chief engineer
West Virginia	Master agreement	West Virginia DOT and Utility Company	The West Virginia DOT has a master agreement that contains procedures the utility and the DOT must follow during the construction of highway projects and establishes the procedures and guidelines to be used when reimbursement of relocation costs is required by law. When the DOT requests a utility to relocate facilities that conflict with highway construction, the utility prepares detailed plans and estimates showing the work to be performed to relocate its facilities, including temporary relocation, if necessary, for both existing and proposed facilities. After the DOT approves these plans and supporting data, the utility may proceed within a reasonable time with the relocation according to the terms and provisions of the master agreement.	Utility owner's authorized representative and WVDOT's deputy State highway engineer for development

APPENDIX D

Sample Conflict Resolution Matrix

Anticipated % of Conflicts to be Resolved at This Level	Time Available for This Level to Resolve Conflict	Department of Transportation	Utility Company
1%	2 weeks	Chief Engineer	CEO or Vice President
9%	10 days	Central Office Directors of Preconstruction and Construction	Utility Project Manager
30%	1 week	Central Office Statewide Utilities Manager	Utility Project Manager
60%	3 days	District Office Resident Engineer and Utilities Coordinator	Utility Field Engineer

We the undersigned agree to make a good-faith effort to undertake and implement this conflict resolution process.

_____

Commissioner, State DOT

_____

CEO, Utility Company

www.ingramcontent.com/pod-product-compliance
Lightning Source LLC
Chambersburg PA
CBHW081624170526
45166CB00009B/3095